# Student Study Guide & Selected Solutions Manual


DAVID REID

*Eastern Michigan University*

JAMES S. WALKER

Upper Saddle River, NJ 07458

Associate Editor: Christian Botting
Senior Editor: Erik Fahlgren
Editor-in-Chief, Physical Sciences: John Challice
Vice President of Production & Manufacturing: David W. Riccardi
Executive Managing Editor: Kathleen Schiaparelli
Assistant Managing Editor: Becca Richter
Production Editor: Dana Dunn
Supplement Cover Management/Design: Paul Gourhan
Manufacturing Buyer: Ilene Kahn
*Cover Photographs: Water droplets - Steve Satushek/Getty Images, Toronto Towers - J.A. Kraulis/Masterfile, Penguins - Art Wolfe/Getty Images, Volcano - Art Wolfe/Getty Images*

Pearson Prentice Hall
Pearson Education, Inc.
Upper Saddle River, NJ 07458



Printed in the United States of America

10 9 8 7 6 5 4 3 2

ISBN 0-13-140653-1

Pearson Education Ltd., *London*
Pearson Education Australia Pty. Ltd., *Sydney*
Pearson Education Singapore, Pte. Ltd.
Pearson Education North Asia Ltd., *Hong Kong*
Pearson Education Canada, Inc., *Toronto*
Pearson Educación de Mexico, S.A. de C.V.
Pearson Education—Japan, *Tokyo*
Pearson Education Malaysia, Pte. Ltd.
Pearson Education, *Upper Saddle River, New Jersey*

## TABLE OF CONTENTS

## PREFACE

This study guide is designed to assist you in your study of the fascinating and sometimes challenging world of physics using *Physics*, *Second Edition* by James S. Walker. To do this I have provided a Chapter Review, which consists of a comprehensive (but brief) review of almost every section in the text. Numerous solved examples and exercises appear throughout each Chapter Review. The examples follow the wonderful two-column format of the text, while the solutions to the exercises have a more traditional layout. Together with the Chapter Review, each chapter contains a list of objectives, a practice quiz, a glossary of key terms and phrases, a table of important formulas, and a table that reviews the units of the new quantities introduced.

In addition to the above materials that I have provided, you will also find Warm-Up and Puzzle questions by Just in Time Teaching innovators Gregory Novak and Andrew Gavrin (Indiana University-Purdue University, Indianapolis), Practice Problems by Carl Adler (East Carolina University) and Solution to Selected Problems from the *Instructor's Solutions Manual*. Taken together, the information in this study guide, when used in conjunction with the main text, should enhance your ability to master the many concepts and skills needed to understand physics, and therefore, the world around you. Work hard, and most importantly, have fun doing it!

I am indebted to many for helping me to complete this work. Most directly, I thank Mr. Christian Botting of Prentice Hall for his continued work with me on this project. I most especially wish the acknowledge Dr. Anand P. Batra of Howard University. He provided an excellent review of the physics content for the first edition of this Study Guide and made countless valuable suggestions.

David D. Reid
Eastern Michigan University
May, 2003

To my wife, *Annie*

# CHAPTER 1

# INTRODUCTION

## Chapter Objectives

After studying this chapter, you should

**1.** know the three most common basic physical quantities in physics and their units.

**2.** know how to determine the dimension of a quantity and perform a dimensional check on any equation.

**3.** be familiar with the most common metric prefixes.

**4.** be able to perform calculations, keeping a proper account of significant figures.

**5.** be able to convert quantities from one set of units to another.

**6.** be able to perform quick order-of-magnitude calculations.

## Warm-Ups

**1.** Do you believe that the metric system is superior to the previous systems of measure, such as the everyday system used in the United States? Whatever your answer, what arguments would you use to persuade a person who has a different opinion?

**2.** Estimate the number of seconds in a human lifetime. We'll let you choose the definition of lifetime. Do all reasonable choices of lifetime give answers that have the same order of magnitude?

**3.** Which is a faster speed, 30 mi/h or 13 m/s? Describe in words how you obtained your answer.

**4.** Estimate how many 20-cm × 20-cm tiles it would take to tile the floor and three sides of a shower stall. The stall has a 16-$ft^2$ floor and 5-ft walls.

## Chapter Review

### 1–1 – 1–2 Physics and the Laws of Nature & Units of Length, Mass, and Time

The study of **physics** deals with the fundamental laws of nature and many of their applications. These laws govern the behavior of all physical phenomena. We describe the behavior of physical systems using various quantities that we create for this purpose; however, there are three quantities—**length**, **mass**, and **time**—that we take as fundamental quantities, and we use these three to create other quantities.

We define a system of units for these quantities so that we can specify how much length, mass, or time we have. The system of units used in this book is the SI, which stands for Système International. In this system, the unit of length is the **meter** (m), the unit of mass is the **kilogram** (kg), and the unit of time is the **second** (s). This system of units is still sometimes referred to by its former name, the mks system.

SI units are based on the metric system. An important aspect of this system is its hierarchy of prefixes used for quantities of different magnitudes. Certain of these prefixes are used very frequently in physics, so you should become very familiar with them. Some of the more common ones are listed here:

| Power | Prefix | Symbol |
|---|---|---|
| $10^{-15}$ | femto | f |
| $10^{-12}$ | pico | p |
| $10^{-9}$ | nano | n |
| $10^{-6}$ | micro | μ |
| $10^{-3}$ | milli | m |
| $10^{-2}$ | centi | c |
| $10^{3}$ | kilo | k |
| $10^{6}$ | mega | M |

**Exercise 1–1 Metric Prefixes** Write the following quantities using a convenient metric prefix.
**(a)** 0.00025 m **(b)** 25,000 m **(c)** 250 m **(d)** 250,000,000 m **(e)** 0.0000025 m

**Solution** **(a)** 0.25 mm **(b)** 25 km **(c)** 0.25 km **(d)** 250 Mm **(e)** 2.5 μm

## Practice Quiz

**1.** Which of the following quantities is not one of the fundamental quantities?

**(a)** length **(b)** speed **(c)** time **(d)** mass

## 1–3 Dimensional Analysis

In physics we derive the physical quantities of interest from the set of fundamental quantities of length, mass, and time. The **dimension** of a quantity tells us what *type* of quantity it is. When indicating the dimension of a quantity only, we use capital letters enclosed in brackets. Thus, the dimension of length is represented by [L], mass by [M], and time by [T].

We use many equations in physics, and these equations must be dimensionally consistent. It is extremely useful to perform a dimensional analysis on any equation about which you are unsure. If the equation is not dimensionally consistent, it cannot be a correct equation. The rules are simple:

* Two quantities can be added or subtracted only if they are of the same dimension.
* Two quantities can be equal only if they are of the same dimension.

Notice that only the dimension needs to be the same, not the units. It is perfectly valid to write 12 inches = 1 foot because both quantities are lengths, [L] = [L], even though their units are different. However, it is not valid to write $x$ inches = $t$ seconds because the quantities have different dimensions: [L] ≠ [T].

---

**Example 1–2 Checking the Dimensions** Given that the quantities $x$ (m), $v$ (m/s), $a$ (m/s$^2$), and $t$ (s) are measured in the units shown in parentheses, perform a dimensional analysis on the following equations. **(a)** $x = t$ **(b)** $x = 2vt$ **(c)** $v = at + t/x$ **(d)** $x = vt + 3at^2$

**Picture the Problem** There is no picture.

**Strategy** Write each equation in terms of its dimensions and check if the equation obeys the preceding rules.

**Solution**

**Part (a)**

**1.** Write the equation with dimensions only:

$$[L] = [T]$$

Because these dimensions are not the same, the equation is not valid.

**Part (b)**

**2.** Write out the dimensions of this equation:

$$[L] = \frac{[L]}{[T]} \times [T] = [L]$$

The right-hand-side dimension is equal to the dimension on the left, so the equation is dimensionally correct.

**Part (c)**

**3.** Write out the dimension of this equation:

$$\frac{[L]}{[T]} = \frac{[L]}{[T^2]} \times [T] + \frac{[T]}{[L]} = \frac{[L]}{[T]} + \frac{[T]}{[L]}$$

The first and second terms on the right-hand side are not of equal dimension and cannot be added. This is not a valid equation.

**Part (d)**

**4.** Write out the dimension of this equation:

$$[L] = \frac{[L]}{[T]} \times [T] + \frac{[L]}{[T^2]} \times [T^2] = [L] + [L]$$

Here, both terms on the right have the same dimension, which is also equal to the dimension on the left. This equation is dimensionally correct.

**Insight** Notice that in dimensional analysis purely numerical factors are ignored because they are **dimensionless**. Because there are dimensionless quantities, dimensional consistency does not guarantee that the equation is physically correct, but it makes for a quick and easy first check.

---

## Practice Quiz

**2.** Which of the following expressions is dimensionally correct?

**(a)** [L] = [M] × [T] **(b)** [T] = [L]/[T] **(c)** $[\mathrm{L}] = \frac{[\mathrm{L}]}{[\mathrm{T}]} \times [\mathrm{T}]$ **(d)** $[\mathrm{M}] = \frac{[\mathrm{L}^2]}{[\mathrm{T}]}$

**3.** If speed $v$ has units of m/s, distance $d$ has units of m, and time $t$ has units of s, which of the following expressions is dimensionally correct?

**(a)** $v = t/d$ **(b)** $t = vd$ **(c)** $d = v/t$ **(d)** $t = d/v$

## 1–4 Significant Figures

All measured quantities carry some uncertainty in their values. When working with the values of quantities, it is important to keep proper account of the digits that are reliably known. Such digits are called **significant figures.** The rules for working with significant figures are as follows:

* *Multiplication and Division*: The number of significant figures in the result of a multiplication or division equals the number of significant figures in the factor containing the fewest significant figures.
* *Addition and Subtraction*: The significant figures in the result of an addition or subtraction are located only in the *places* (hundreds, ones, tenths, etc.) that are reliably known for *every* value in the sum.

---

**Example 1–3 Driving in a Residential Zone** On most residential streets in the United States the speed limit is 25 mi/h (= 11 m/s). If a car drives down a neighborhood side street at the legal speed limit for 120.46 s, how much distance does the car cover?

**Picture the Problem** Our sketch shows the car moving along a straight road.

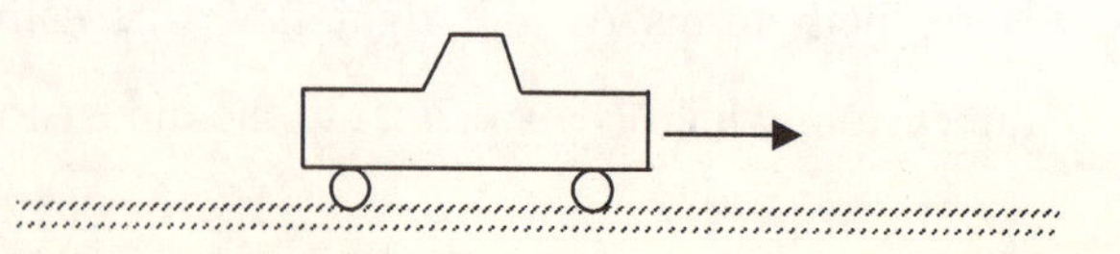

**Strategy** The distance traveled is the speed multiplied by the time of travel.

**Solution**

Multiply the speed and the time to get the distance: $d = 11 \text{ m/s} \times 120.46 \text{ s} = 1300 \text{ m}$

**Insight** There are two important things to notice about the result. First, despite the fact that the time is known to five significant figures, the speed is known only to two, and so the result has only two significant figures. Second, the final two zeros in the value 1300 are not significant. They must be written, however, to give the proper magnitude of the value. It can often be unclear whether such zeros are significant. This problem can be avoided by using **scientific notation** (discussed later).

---

**Exercise 1–4 Significant Figures** A calculation involves the addition of two measured distances $d_1$ = 1250 m, and $d_2$ = 336 m. If each measurement is given to three significant figures, what is the result of the calculation?

**Solution** Adding the two distances we get $d_1 + d_2 = 1250 \text{ m} + 336 \text{ m} = 1590 \text{ m}$.

The answer is not 1586 because even though the 6 is significant in 336 m, the one's place of 1250 m (the 0) is not significant, so the one's place of the result cannot be significant. You may wonder about the fact that there is no significant figure in the thousand's place of 336 m because the value requires no digit there; however, because there is no digit there, we know that place with certainty.

---

**Round-off Error**

Be aware that to avoid excessive round-off error you should round only to the proper number of significant figures at the very end of a calculation. In Example 1–4, if the distance calculated is only an intermediate step in a longer calculation, then the value 1586 m should be used in the subsequent steps. In general, keep at least one additional digit for values calculated in intermediate steps. Another, even better, approach is not to calculate intermediate values numerically, but to just carry through the formulas inserting numerical values only at the end.

**Example 1–5 Don't Round-off too Soon** A cardboard box has measurements of $L$ = 1.92 m, and $W$ = 0.725 m. Its height is $H$ = 1.88 m. **(a)** Calculate the area ($A$) of the base of the box. **(b)** Calculate its volume ($V$) using the result of (a). **(c)** Calculate its volume using the formula for volume.

**Picture the Problem** The diagram shows a box representing the box whose base area and volume we wish to determine.

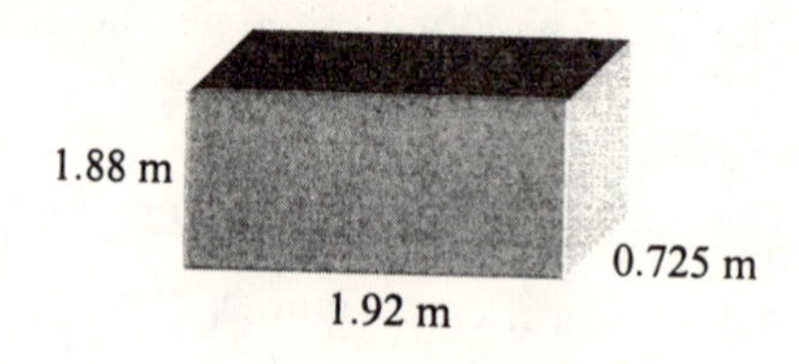


**Strategy** First calculate the area of the base:

**Solution**

**1.** Calculate the area of the base:

$A = LW = 1.92\text{ m} \times 0.725\text{ m} = 1.39\text{ m}^2$

**2.** The volume of the box is area × height:

$V = A \times H = 1.39\text{ m}^2 \times 1.88\text{ m} = 2.61\text{ m}^3$

**3.** The volume is length × width × height:

$V = A \times H = LWH = 1.92\text{ m} \times 0.725\text{ m} \times 1.88\text{ m}$

$= 2.62\text{ m}^3$

**Insight** The answers to parts (b) and (c) differ in the final digit. Which one is correct? Part (c) is correct because the full values were used. The round-off to three significant figures in part (a) is the reason for the difference.

---

**Scientific Notation**

A very useful way of writing numerical values is to use scientific notation. In this notation, a value is written as a number of order unity (meaning that only one digit is left of the decimal point) times the appropriate power of 10. The value of scientific notation is that it allows for quick identification of the **order-of-magnitude** (power of 10) of a quantity, calculations are often easier to perform when the values are listed this way, and it removes any ambiguity in the number of significant figures. For example, if the number 3500 has only one significant figure, we write it as $4 \times 10^3$; if it has two, we write $3.5 \times 10^3$; if it has three, we write $3.50 \times 10^3$; and if it has four significant figures we write $3.500 \times 10^3$.

---

**Exercise 1–6 Scientific Notation** Write the following quantities using scientific notation assuming three significant figures.

**(a)** 0.00250 m **(b)** 12,060 m **(c)** 451 m **(d)** 8.00 m **(e)** 0.00003593 m

**Answer:** **(a)** $2.50 \times 10^{-3}$ m **(b)** $1.21 \times 10^{4}$ m **(c)** $4.51 \times 10^{2}$ m **(d)** $8.00 \times 10^{0}$ m **(e)** $3.59 \times 10^{-5}$ m

When the value is already of order unity, as with part (d), the power of 10 is often dropped. In such cases, the fact that the two zeros are written after the decimal point indicates that they are significant figures.

---

**Practice Quiz**

**4.** Assuming that every nonzero digit is significant, consider the following product of numbers: $1.34 \times 10.75 \times 0.042$. Which answer is correct to the proper number of significant figures?

**(a)** 6 **(b)** 0.61 **(c)** 6.05 **(d)** 6.0501

**5.** Assuming that only nonzero digits are significant, consider the following sum of numbers: 1700 + 338 + 13. Which answer is correct to the proper number of significant figures?

**(a)** 2051 **(b)** 2050 **(c)** 2100 **(d)** 2000

**6.** Consider the following expression: $(5.93) \times (8.762) + (2.116) \times (3.70)$. Which answer is correct to the proper number of significant figures?

**(a)** 59.78786 **(b)** 59.79 **(c)** 59.83 **(d)** 59.8

**7.** Which of the following numbers is the proper scientific notation for 25,300?

**(a)** $2.53 \times 10^{4}$ **(b)** $25.3 \times 10^{3}$ **(c)** $2.53 \times 10^{3}$ **(d)** $0.253 \times 10^{5}$

**8.** The number $7.4 \times 10^{5}$ is equivalent to which of the following?

**(a)** 7.4 **(b)** 740 **(c)** 7,400 **(d)** 740,000

## 1–5 Converting Units

Even though we predominantly use SI units, it will often be necessary to convert between SI and other units. A conversion can be accomplished using a **conversion factor** that is constructed by knowing how much of a quantity in one unit equals that same quantity in another unit. A conversion factor is a ratio of equal quantities written such that, when multiplied by a quantity, the undesired unit algebraically cancels leaving only the desired unit. This concept is best illustrated by example.

---

**Example 1–7 Volume of a Box** A typical cardboard box provided by moving companies measures 1.50 ft × 1.50 ft × 1.33 ft. Determine the volume ($V$) of clothes that you can pack into this box in cubic meters.

**Picture the Problem** The diagram represents the box whose volume we wish to determine.

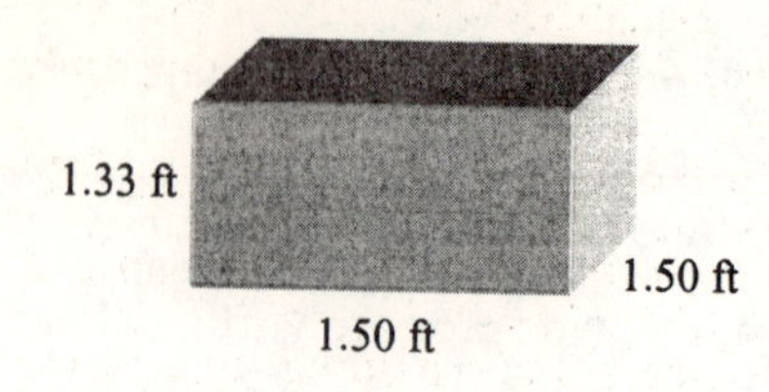


**Strategy** We first calculate the volume in the given units, determine the conversion factor, and then convert the volume to cubic meters.

**Solution**

**1.** Calculate the volume as given: $V = 1.50\ \text{ft} \times 1.50\ \text{ft} \times 1.33\ \text{ft} = 2.993\ \text{ft}^3$

**2.** Write the number of meters in a foot: $1\ \text{m} = 3.281\ \text{ft}$

**3.** Write the conversion factor from feet to meters: $\dfrac{1\ \text{m}}{3.281\ \text{ft}}$

**4.** The conversion factor from $\text{ft}^3$ to $\text{m}^3$ is: $\left(\dfrac{1\ \text{m}}{3.281\ \text{ft}}\right)^3 = \dfrac{1\ \text{m}^3}{35.32\ \text{ft}^3}$

**5.** Multiply the volume by the conversion factor: $V = 2.993\ \text{ft}^3 \left(\dfrac{1\ \text{m}^3}{35.32\ \text{ft}^3}\right) = 0.0847\ \text{m}^3$

**Insight** In the final step, the unit $\text{ft}^3$ cancels just as numbers would. Setting up this cancellation is the crucial step in unit conversion. You will get plenty of practice converting units in your study of physics.

---

**Practice Quiz**

9. Given that 1 in. = 2.54 cm, convert 250.0 cm to inches.

**(a)** 635 in. **(b)** 0.394 in. **(c)** 98.4 in. **(d)** 150.0 in.

10. Convert the speed $1.00 \times 10^2$ m/s to km/h.

**(a)** 36 km/h **(b)** 360 km/h **(c)** $3.60 \times 10^8$ km/h **(d)** 27.8 km/h

## 1–7 Problem Solving in Physics

Solving physics problems is a logical and creative endeavor for which there is no set prescription; however, there are certain practices that help this creative process to flourish. First, a *careful reading* of the problem is necessary to fully grasp the question being posed and the information being given. It is often useful to separately write out all the given and required information; several of the solved examples in this study guide illustrate that approach. It is also a good practice to make a *sketch* of the problem and to *visualize* the physics that is taking place. A correct mental picture of the problem takes you a long way toward a correct solution. Next, map out your *strategy* for the solution. Here, you basically solve the problem logically before doing it mathematically. For the mathematical solution, you need to *identify and solve* the appropriate equations for the relevant physics. Finally, you should *check and explore* your result to be sure that the answer makes sense in the context of the problem.

## Reference Tools and Resources

### I. Key Terms and Phrases

**physics** the study of the fundamental laws of nature and many of their applications

**SI units** the internationally adopted standard system of units (based on meters, kilograms, and seconds) for quantitatively measuring quantities

**dimension of a quantity** the fundamental type of a quantity such as length, mass, or time

**dimensional analysis** a type of calculation that checks the dimensional consistency of an equation

**significant figures** the digits in the numerical value of a quantity that are known with certainty

**scientific notation** a method of writing numbers that consists of a number of order unity times the appropriate power of 10

**conversion factor** a factor (equal to 1) that multiplies a quantity to convert its value to another unit

**order-of-magnitude** the power of 10 characterizing the size of a quantity

### II. Tips

**Dimensional Analysis**

You should be aware that, typically, arguments of mathematical functions are dimensionless. Angles, for example, are dimensionless, as can be seen by the equation for the length of a circular arc, $s = r\theta$, where $\theta$ is in radians; hence, angular measures such as radians and degrees signify only how we choose to measure the angle. The trigonometric functions, therefore, such as sine, cosine, and tangent are applied to dimensionless quantities. Other examples of dimensionless functions are log($x$), ln($x$), and their inverse functions $10^x$ and exp($x$).

## Practice Problems

1. What is the decimal equivalent of $3.14 \times 10^3$?
2. What is the decimal equivalent of 2.7e–4?

   (In the JavaScript web language, scientific notation normally written as $2 \times 10^3$ is written as 2e3.)

3. 10.2 * 7.4 =
4. 27.1/5.09 =
5. 2.712 + 10.1 =
6. What is the volume in cubic centimeters of a sphere with a radius of 2 cm?
7. A sphere has a volume of 106 $cm^3$; what is its radius (in cm)?
8. What is the area in square meters of a triangle with a base 2.8 m and a height of 10.7 m?
9. What is the volume in cubic centimeters of a cylinder of diameter 1.7 cm with a height of 2.4 cm?
10. 26 miles (exactly) is how many meters (exactly)?

## Puzzle

**WHERE ARE YOU?**

The standard geographical coordinates of Chicago are as follows:

Latitude: 41 degrees 50 minutes

Longitude: 87 degrees 45 minutes.

What are the $x$, $y$, $z$ coordinates of Chicago in a coordinate system centered at the center of Earth, with the $z$-axis pointing from the South Pole to the North Pole, and the $x$-axis passing through the zero longitude meridian pointing away from Europe into space? Answer this question in words, not equations, briefly explaining how you obtained your answer.

## Selected Solutions

**9.** We first solve the equation for $k$:

$$T = 2\pi\sqrt{\frac{m}{k}} \quad \Rightarrow \quad T^2 = 4\pi^2\frac{m}{k} \quad \Rightarrow \quad k = 4\pi^2\frac{m}{T^2}.$$

The factor $4\pi^2$ is dimensionless, so dimensional analysis gives $\boxed{[\mathrm{k}] = \frac{[\mathrm{M}]}{[\mathrm{T}^2]}}$

**13.** The total weight is given by the sum

$$Wt_{\text{tot}} = Wt_{\text{bass}} + Wt_{\text{cod}} + Wt_{\text{salmon}}$$

Taking the direct sum we obtain

$$Wt_{\text{tot}} = 2.45\ \text{lb} + 10.1\ \text{lb} + 16.13\ \text{lb} = 28.68\ \text{lb}$$

This result must be rounded to the proper number of significant figures. The 10.1-lb cod gives us only one decimal place. Therefore, the result must be rounded to one decimal place:

$$Wt_{\text{tot}} = \boxed{28.7\ \text{lb}}$$

**15.** The area of a circle is given by the expression $A = \pi r^2$. The value of $\pi$ is known to many significant digits. The number of significant digits in the results is limited only by the number in the radius $r$. In each case, we need to use only one more significant digit in $\pi$ than in $r$ to obtain the most accurate result available.

**(a)** $A = (3.1416)(5.142 \text{ m})^2 = \boxed{83.06 \text{ m}^2}$

**(b)** $A = (3.14)(1.7 \text{ m})^2 = \boxed{9.1 \text{ m}^2}$

**29.** **(a)** For this part, we need only to convert meters to feet:

$$\left(\frac{20.0 \text{ m}}{\text{s}}\right)\left(\frac{3.28 \text{ ft}}{\text{m}}\right) = \boxed{65.6 \text{ ft/s}}$$

**(b)** For this part, we must convert both the distance and time units.

$$\left(\frac{65.6 \text{ ft}}{\text{s}}\right)\left(\frac{1 \text{ mi}}{5280 \text{ ft}}\right)\left(\frac{3600 \text{ s}}{\text{h}}\right) = \boxed{44.7 \text{ mi/h}}$$

**33.** **(a)** From the given information, the surface of Earth must move through the 3000 mi in the 3-h time difference. Thus, we have for the rotational speed,

$$v = \frac{3000 \text{ mi}}{3 \text{ h}} = \boxed{1000 \text{ mi/h}}$$

**(b)** Earth completes one rotation in 24 h. Given the above rotational speed, we can say

$$\frac{1000 \text{ mi}}{1 \text{ h}} = \frac{\text{Circum.}}{24 \text{ h}} \quad \therefore \quad \text{Circum.} = 24 \text{ h} \times \frac{1000 \text{ mi}}{\text{h}} = \boxed{24{,}000 \text{ mi}} \quad \text{(approx.)}$$

**(c)** The circumference of a circle is given by $\text{Circum.} = 2\pi r$. Therefore,

$$r = \frac{\text{Circum.}}{2\pi} = \frac{24{,}000 \text{ mi}}{2\pi} = \boxed{4000 \text{ mi}}$$

**Answers to Practice Quiz**

**1.** (b) **2.** (c) **3.** (d) **4.** (b) **5.** (c) **6.** (d) **7.** (a) **8.** (d) **9.** (c) **10.** (b)

**Answers to Practice Problems**

**1.** 3,140 **2.** 0.00027

**3.** 75 **4.** 5.32

**5.** 12.8 **6.** $30 \text{ cm}^3$

**7.** 2.94 cm **8.** $15 \text{ m}^2$

**9.** $5.4 \text{ cm}^3$ **10.** 41,842.944 m

# CHAPTER 2

# ONE-DIMENSIONAL KINEMATICS

## Chapter Objectives

After studying this chapter, you should

1. know the difference between distance and displacement.
2. know the difference between speed and velocity.
3. know the difference between velocity and acceleration.
4. be able to define acceleration and give examples of both positive and negative acceleration.
5. be able to calculate displacements, velocities, and accelerations using the equations of one-dimensional motion.
6. be able to interpret $x$-versus-$t$ and $v$-versus-$t$ plots for motion with both constant velocity and constant acceleration.
7. be able to describe the motion of freely falling objects.

## Warm-Ups

1. During aerobic exercise, people often suffer injuries to knees and other joints due to high accelerations. When do these high accelerations occur?
2. Estimate the acceleration you subject yourself to if you walk into a brick wall at normal walking speed. (Make a reasonable estimate of your speed and of the time it takes you to come to a stop.)
3. A man drops a baseball from the edge of the roof of a building. At exactly the same time, another man shoots a baseball vertically up toward the man on the roof in such a way that the ball just barely reaches the roof. Does the ball from the roof reach the ground before the ball from the ground reaches the roof, or vice versa?
4. Estimate the time it takes for a free-fall drop from a height of 10 m. Also estimate the time a 10-m platform diver is in the air if he takes off straight up with a vertical speed of 2 m/s (and clears the platform of course!)

## Chapter Review

### 2–1 Position, Distance, and Displacement

Any description of motion takes place in a **coordinate system** that allows us to track the **position** of an object. **One-dimensional motion** means that objects are free to move back and forth only along a single